A Letter to Parris Darwin

Parris,

From the moment you were born, I knew there was something special about you. You have a bright mind, a kind heart, and a curiosity that can take you anywhere in the world

This book began as a fun adventure about a little water droplet named William, but it became something much bigger. It became a reminder that every journey in life matters – just like every drop of water matters.

As you grow older, I want you to remember something important. You can become anything you dream of becoming. Whether you choose to be an engineer, scientist, athlete, artist, leader, or something the world has never seen before, never stop believing in yourself. The world needs young people who are curious, brave, thoughtful, and willing to help others. Just like water keeps life moving, your ideas, your voice, and your actions can help make the world a better place.

Always stay humble.
Always keep learning.
Always protect what matters.
And never be afraid to dream big.
I hope this story reminds you that even the smallest things can make a huge difference.

And no matter where life takes you, keep exploring, keep asking questions, and keep believing that your journey matters too.

With love and pride.
Uncle Man-Man.

Hi! My name is Parris Darwin.
I love to explore, ask big questions, and learn how the world works—especially water!

And I'm William the Water Droplet!
I travel all around the Earth in ways
you might not expect.

One day, Parris asked,
"William... where do you come from?"

"When the sun warms me up,"
said William,
"I rise into the sky as vapor.
This is called evaporation!"

"High in the sky, I cool down and join
other droplets to form clouds.
That's called condensation."

"When clouds get heavy, we fall back to Earth as rain, snow, or hail. That's precipitation!"

Parris smiled. “Wow! Where do you go next?”

"Sometimes I flow through rivers and streams,
traveling across the land with other droplets."

"Other times, I soak into the ground.
This is called infiltration."

"Underground, I travel through spaces in soil and rock called aquifers."

"Many of us make our way back to the ocean…
and then the journey begins all over again!"

"This never-ending journey,"
said William,
"is called the hydrologic cycle."

Water is always moving—
from sky to land to water
and back again!

"Clean water helps people, plants, and animals stay healthy," said Parris. "That sounds important!"

"What do you think happens if water gets dirty?"

William nodded.
"That's why there are people whose job is to protect water every day!"

ENVIRONMENTAL ENGINEER
"Environmental Engineers design systems to keep water clean and safe."
They can earn:
$75,000 – $140,000+
With experience, these roles can earn even more!
WATER SOLUTIONS
ENVIRONMENTAL ENGINEER
PD
WATER SYSTEMS DESIGN
RAIN
COLLECTION
TREATMENT
STORAGE
DISTRIBUTION
PLAN TODAY PROTECT TOMORROW
DESIGN SOLVE PROTECT
WATER TREATMENT SYSTEM
INTAKE → SCREENING → TREATMENT → STORAGE → CLEAN WATER
Clean Water Healthy Communities
WATER SYSTEMS
SUSTAINABLE SOLUTIONS
ENGINEERING A BETTER FUTURE

WATER OPERATOR

"Water Operators help run systems that treat and deliver clean water to communities."

$60,000 – $120,000+

⭐ Some operators start with training and certifications instead of a 4-year college degree.

ENVIRONMENTAL INSPECTOR (WATER QUALITY)

"Environmental Inspectors check sites and water systems to make sure water stays clean and safe."

They can earn:
$70,000 – $120,000+
(Top roles can exceed $130,000)

Some roles can be started with a high school diploma and certifications.

"You can start learning about water in school, training programs, and hands-on experiences." Said William.

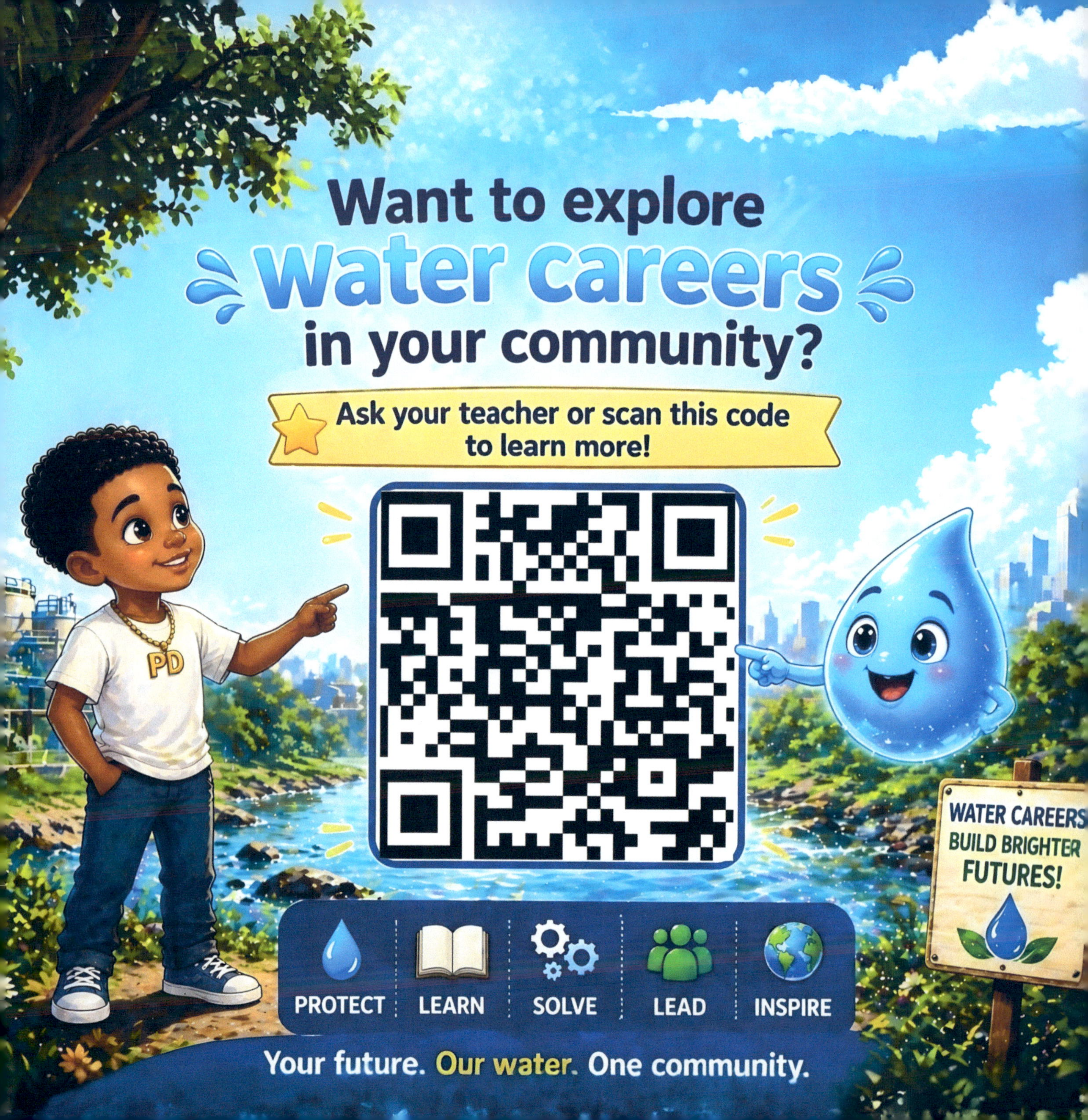
Want to explore
water careers
in your community?
Ask your teacher or scan this code
to learn more!
PD
WATER CAREERS
BUILD BRIGHTER
FUTURES!
PROTECT
LEARN
SOLVE
LEAD
INSPIRE
Your future. Our water. One community.

Parris smiled.
"I didn't know water had such a big journey.
Maybe one day, I can help protect it too!"

The adventure continues...
Next stop: Groundwater!
Thank you again.

About the Author

Vidal Daniels is an environmental professional, educator, and community advocate dedicated to inspiring the next generation to protect our world and explore careers in sustainability and water. He has worked in both the public and private sectors, supporting environmental programs related to stormwater, wastewater, water quality, and regulatory compliance. Vidal has served in municipal government with the City of Modesto and the City of Brentwood, and has also worked as an environmental consultant with WSP, helping support environmental inspections, infrastructure initiatives, and sustainable community solutions.

Vidal holds multiple environmental and safety certifications, including Environmental Compliance Inspector II (CWEA), Qualified Industrial Stormwater Practitioner (QISP), Qualified SWPPP Developer/Practitioner (QSP/D), Certified Professional in Erosion and Sediment Control (CPESC), OSHA 510 Construction Safety certification, HAZWOPER certification, and additional environmental compliance credentials. He earned an undergraduate degree in Biological Sciences and is currently completing a Master's degree in Environmental Engineering. Beyond his professional work, Vidal is passionate about workforce development, environmental education, and helping young people discover meaningful career pathways. Through DFT Solutions Group LLC and his community involvement with organizations such as USGBC California and 100 Black Men of America (Sacramento Chapter), he works to connect education, sustainability, and opportunity in underserved communities. "William the Water Droplet & the Adventures of Parris Darwin" was created to make learning about water fun, inspiring, and accessible for children everywhere — while encouraging future leaders, scientists, engineers, and environmental stewards to dream big and help protect our planet.

www.ingramcontent.com/pod-product-compliance
Lightning Source LLC
LaVergne TN
LVHW071239210926
848486LV00007B/301
* 9 7 9 8 9 9 6 9 6 6 6 9 1 *